Electronic Configuration: A Formula Handbook

N.B. Singh

DEDICATION

To Nature,

I dedicate this book to you, the source of all life. You are my inspiration, my teacher, and my friend.

Thank you for teaching me about the beauty of the world around me. Thank you for showing me the power of the natural world. Thank you for giving me a sense of peace and tranquillity.

I promise to do my part to protect you and your many wonders. I will teach my children about the importance of conservation and sustainability. I will work to make the world a better place for all living things.

Thank you for everything, Nature.

With love,

N.B Singh

Contents

9   Experimental Techniques   45

10   Environmental Applications   49

Preface

Welcome to *Electronic Configuration: A Formula Handbook*. This book aims to provide a comprehensive guide to understanding and applying electronic configurations in the field of chemistry. Whether you are a student, researcher, or enthusiast, this handbook is designed to be a valuable resource.

Purpose of the Handbook

The primary goal of this handbook is to present electronic configurations in a clear and concise manner. We cover a wide range of topics, from the basics of electron distribution to advanced applications in computational chemistry. The content is structured to facilitate easy navigation, making it suitable for both learning and quick reference.

Organization of the Book

The book is divided into several chapters, each focusing on a specific aspect of electronic configuration. Starting with the fundamentals, we gradually delve into more advanced topics. The chapters are organized logically to build a strong foundation and facilitate progressive learning.

Features of the Handbook

- Clear explanations of electronic configurations using both traditional notation and modern computational methods.

- Practical examples and case studies to reinforce concepts.

- Comprehensive coverage of relevant topics in the field of chemistry.

Who Should Read This Book

This handbook is intended for students studying chemistry, researchers working in the field, and anyone interested in gaining a deeper understanding of electronic configurations. Whether you are new to the subject or seeking a quick reference guide, you will find valuable insights within these pages.

Thank you for choosing *Electronic Configuration: A Formula Handbook*. We hope you find it informative and enjoyable.

Chapter 1

Introduction

1.1 Overview

In this section, we delve into the foundational concepts that shape the understanding of electronic configurations. The exploration of atomic structure, quantum mechanics, and electron configurations forms the basis for comprehending the behavior of atoms in chemistry.

1.1.1 Atomic Structure

The atom is composed of a nucleus, housing protons and neutrons, orbited by electrons. Quantum mechanics elucidates the dual nature of electrons, described by the Schrödinger equation and quantum numbers. These numbers (n, l, m, s) delineate the quantum state of electrons.

1.1.2 Quantum Mechanics

Wave-particle duality characterizes quantum mechanics, allowing electrons to exhibit both particle and wave properties. The Schrödinger equation governs the behavior of electrons within an atom, and quantum numbers provide insights into their spatial distribution.

1.1.3 Electron Configurations

Understanding how electrons are distributed within an atom is crucial. The Aufbau principle, Pauli exclusion principle, and Hund's rule guide the arrangement of electrons in orbitals. This forms the foundation for predicting chemical properties.

1.1.4 Notations and Conventions

Various notations, including orbital diagrams and noble gas notation, simplify the representation of electron configurations. These visual aids are essential for conveying complex electronic structures efficiently.

1.1.5 Applications in Chemistry

Electronic configurations play a pivotal role in explaining chemical bonding and periodic trends. For example, the electron configuration of carbon (C) ($1s^2 2s^2 2p^2$) elucidates its ability to form four covalent bonds.

1.1.6 Advanced Topics

Magnetic properties and electronic transitions broaden our understanding of electronic configurations. Magnetic behavior is influenced by unpaired electrons, and electronic transitions explain phenomena such as absorption and emission spectra.

1.1.7 Modern Techniques

Quantum chemistry and computational methods provide powerful tools for predicting electronic configurations. Experimental techniques like X-ray crystallography offer direct insights into the spatial arrangement of electrons.

1.1.8 Chemical Reactions and Bonding

Chemical reactions involve the rearrangement of electrons. Understanding electron configurations is crucial for predicting the outcomes of reactions. Bonding diagrams, such as those for H_2 and O_2, showcase the sharing and pairing of electrons.

1.1.9 Numerical Example

Consider the electron configuration of nitrogen (N). It is $1s^2 2s^2 2p^3$, indicating three unpaired electrons in the 2p orbital. This configuration explains nitrogen's ability to form three bonds, as seen in NH_3.

1.2 Objectives

The objectives of this introductory chapter are to set the stage for a comprehensive exploration of electronic configurations in the realm of atomic and molecular structures. By the end of this chapter, readers should achieve the following key objectives:

1.2.1 Understand Atomic Structure

Grasp the fundamental components of an atom, including protons, neutrons, and electrons. Comprehend the spatial distribution of electrons in orbitals and the role of quantum numbers in describing their states.

1.2.2 Appreciate Quantum Mechanics

Develop an appreciation for the principles of quantum mechanics, particularly the wave-particle duality of electrons. Understand how the Schrödinger equation governs the behavior of electrons within the quantum framework.

1.2.3 Master Electron Configurations

Become proficient in determining electron configurations using the Aufbau principle, Pauli exclusion principle, and Hund's rule. Learn how to represent these configurations using various notations and conventions.

1.2.4 Explore Chemical Applications

Recognize the significance of electronic configurations in explaining chemical bonding, periodic trends, and the behavior of elements in various chemical reactions. Gain insights into the role of electrons in forming and breaking chemical bonds.

1.2.5 Utilize Modern Techniques

Familiarize yourself with modern techniques such as quantum chemistry, computational methods, and experimental approaches like X-ray crystallography. Understand how these techniques contribute to our understanding of electronic configurations.

1.2.6 Apply Knowledge to Examples

Apply theoretical knowledge to practical examples, including the electron configurations of specific elements, molecular structures, and chemical reactions. Use `\ce` for chemical formulas and `\chemfig` for visualizing molecular structures in text mode.

1.2.7 Develop Problem-Solving Skills

Enhance problem-solving skills by working through numerical examples that require the determination of electron configurations and the prediction of chemical behavior based on electronic structures.

1.2.8 Visualize Electron Configurations

Use `\usepackage{modiagram}` to visually represent electron configurations in a clear and concise manner. Create diagrams that illustrate the arrangement of electrons in different energy levels and orbitals.

1.2.9 Prepare for Advanced Topics

Prepare a solid foundation for delving into advanced topics such as magnetic properties, electronic transitions, and computational methods. Understand how these concepts extend our understanding of electronic configurations.

1.2.10 Build a Strong Framework

Establish a strong conceptual framework that will serve as the basis for the subsequent chapters in this handbook. Lay the groundwork for a comprehensive exploration of electronic configurations and their implications in the world of chemistry.

This chapter serves as a roadmap, guiding readers through the essential concepts and skills needed to navigate the intricate world of electronic configurations. Subsequent chapters will delve deeper into each objective, providing detailed explanations, working examples, and practical applications.

Chapter 2

Atomic Structure

2.1 Fundamental Concepts

This section explores the fundamental concepts that form the basis of understanding atomic structure—the cornerstone of electronic configurations.

2.1.1 Atomic Components

Atoms are the building blocks of matter, comprising protons, neutrons, and electrons. Protons and neutrons reside in the nucleus, while electrons orbit around the nucleus. The mass of an atom is primarily concentrated in the nucleus, and electrons contribute negligibly to the overall mass.

2.1.2 Quantum Mechanics Overview

The behavior of electrons is described by quantum mechanics, a branch of physics that introduces the concept of wave-particle duality. Electrons exhibit both particle-like and wave-like properties. The Schrödinger equation, a cornerstone of quantum mechanics, mathematically describes the behavior of electrons in an atom.

2.1.3 Wave-Particle Duality

Wave-particle duality is a fundamental principle in quantum mechanics, suggesting that particles like electrons can exhibit both wave and particle characteristics. This duality is essential to understanding the behavior of electrons in atomic orbitals.

2.1.4 Quantum Numbers

Quantum numbers are integral to describing the quantum state of electrons. The principal quantum number (n), azimuthal quantum number (l), magnetic quantum number (m), and spin quantum number (s) collectively define an electron's location, orientation, and spin within an atom.

2.1.5 Spatial Distribution of Electrons

Electrons are distributed in various orbitals based on their quantum numbers. The spatial distribution of electrons is visualized using the `\usepackage{modiagram}`, which provides clear diagrams depicting different energy levels and orbital shapes.

2.1.6 Atomic Orbitals

Atomic orbitals are regions in an atom where electrons are likely to be found. The three main types of orbitals are s, p, and d, each with a specific shape and orientation. The `\chemfig` package allows for the visual representation of these orbitals in text mode.

2.1.7 Electron Spin and Pauli Exclusion Principle

Electron spin is a fundamental property, with electrons having either spin up ($\uparrow$) or spin down ($\downarrow$). The Pauli exclusion principle states that no two electrons in an atom can have the same set of quantum numbers, including spin.

2.1.8 Numerical Example

Consider the electron configuration of carbon (C): $1s^2 2s^2 2p^2$. This configuration illustrates the distribution of electrons in the 1s, 2s, and 2p orbitals, following the principles of quantum mechanics.

2.1.9 Applications in Chemistry

Understanding atomic structure and quantum mechanics is foundational to explaining chemical bonding and reactivity. The arrangement of electrons in orbitals influences an atom's ability to form bonds and participate in chemical reactions.

2.2 Quantum Mechanics

Quantum mechanics provides the theoretical framework for understanding the behavior of electrons within an atom. This section delves into the key principles and equations that govern the quantum realm.

2.2.1 Wave-Particle Duality

Wave-particle duality is a cornerstone of quantum mechanics, asserting that particles such as electrons exhibit both wave-like and particle-like characteristics. The de Broglie wavelength (λ) relates the momentum (p) and the wavelength of a particle through the equation $\lambda = \frac{h}{p}$, where h is the Planck constant.

2.2.2 Schrödinger Equation

The Schrödinger equation is a fundamental equation in quantum mechanics that describes how the quantum state of a physical system changes over time. For an electron in an atom, the time-independent Schrödinger equation is given by:

$$H\Psi = E\Psi$$

where H is the Hamiltonian operator, Ψ is the wave function, and E is the energy eigenvalue.

2.2.3   Quantum Numbers

The quantum numbers (n, l, m, s) provide a comprehensive description of an electron's quantum state. The principal quantum number (n) determines the energy level, the azimuthal quantum number (l) determines the orbital shape, the magnetic quantum number (m) specifies the orbital orientation, and the spin quantum number (s) describes the electron's spin.

2.2.4   Wavefunctions and Orbitals

The wave function (Ψ) represents the probability amplitude of finding an electron in a particular quantum state. Orbitals, mathematical functions derived from solutions to the Schrödinger equation, describe the spatial distribution of electrons in an atom.

2.2.5   Heisenberg Uncertainty Principle

The Heisenberg Uncertainty Principle states that it is impossible to simultaneously know both the precise position and momentum of a particle. Mathematically, it is expressed as $\Delta x \cdot \Delta p \geq \frac{\hbar}{2}$, where $\hbar$ is the reduced Planck constant.

2.2.6   Electron Density

The electron density (ρ) is a representation of the probability density of an electron within a given region of space. It is determined by squaring the magnitude of the wave function $(|\Psi|^2)$.

2.2.7   Numerical Example

Consider the hydrogen atom (H). The quantum numbers for the ground state electron are $n = 1$, $l = 0$, $m = 0$, and $s = \frac{1}{2}$. The corresponding wave function and electron density provide insights into the spatial distribution of the electron.

2.2.8 Applications in Chemistry

Quantum mechanics is foundational to explaining the electronic structure of atoms and molecules, influencing chemical bonding, molecular shapes, and spectroscopy. For instance, the molecular orbital theory employs quantum mechanics to describe the distribution of electrons in molecules.

2.2.9 Modern Developments

Modern developments in quantum mechanics, such as density functional theory (DFT) and computational methods, have further advanced our ability to predict and understand electronic configurations in complex systems.

Chapter 3

Electron Configurations

3.1 Aufbau Principle

The Aufbau principle is a fundamental concept in electron configurations, guiding the order in which electrons fill atomic orbitals. This section explores the principles of the Aufbau process and its implications for understanding the arrangement of electrons within atoms.

3.1.1 Sequential Filling of Orbitals

The Aufbau principle dictates that electrons fill the lowest energy orbitals first before moving to higher energy levels. The order of filling is determined by the increasing order of the principal quantum number (n), with lower n values corresponding to lower energy levels.

3.1.2 Sublevel Filling Order

Within a given energy level, the sublevels (s, p, d, f) have distinct energies, with s being the lowest and f the highest. The Aufbau process follows a specific order: 1s, 2s, 2p, 3s, 3p, 4s, 3d, 4p, and so on. This sequence ensures that orbitals are filled in a way that minimizes the overall energy of the atom.

3.1.3 Pauli Exclusion Principle

The Pauli exclusion principle states that no two electrons in an atom can have the same set of quantum numbers. This principle prevents electrons from occupying the same orbital and ensures the unique distribution of electrons in different atomic orbitals.

3.1.4 Hund's Rule

Hund's rule dictates that electrons fill degenerate orbitals (orbitals with the same energy level) singly before pairing up. This results in unpaired electrons in degenerate orbitals, maximizing the total electron spin and stabilizing the atom.

3.1.5 Numerical Example

Consider the electron configuration of nitrogen (N). Following the Aufbau principle, the order of filling is $1s^2 2s^2 2p^3$, where the first two electrons occupy the 1s orbital, and the remaining three fill the 2s and 2p orbitals.

3.1.6 Applications in Chemistry

The Aufbau principle is foundational to predicting the electronic configurations of elements. It plays a crucial role in explaining the periodic table's structure, periodic trends, and the chemical behavior of elements.

3.1.7 Visualization with `modiagram`

Using the `\usepackage{modiagram}`, the electron configuration of an atom can be visually represented with clear diagrams. For example, the electron configuration of nitrogen (N) can be depicted using the following code:

$$1s^2 2s^2 2p^3$$

This diagram visually represents the filling of the 1s and 2s orbitals, followed by the filling of the 2p orbitals in nitrogen.

3.2 Pauli Exclusion Principle

The Pauli Exclusion Principle is a fundamental concept in quantum mechanics, stating that no two electrons in an atom can have the same set of quantum numbers. This section explores the significance of the Pauli Exclusion Principle in electron configurations.

3.2.1 Unique Quantum States

The Pauli Exclusion Principle ensures that each electron within an atom occupies a unique quantum state characterized by a specific set of quantum numbers. These quantum numbers include the principal quantum number (n), azimuthal quantum number (l), magnetic quantum number (m), and spin quantum number (s).

3.2.2 Spin Quantum Number

The spin quantum number (s) is a crucial component of the Pauli Exclusion Principle. It has two possible values: $+\frac{1}{2}$ (spin up) and $-\frac{1}{2}$ (spin down). Electrons in the same orbital must have opposite spins, ensuring their distinct quantum states.

3.2.3 Matrix Representation

The Pauli Exclusion Principle can be represented using a matrix for electron spin states. For example, the matrix representation for two electrons in an orbital is:

$$\begin{bmatrix} \uparrow & 0 \\ 0 & \downarrow \end{bmatrix}$$

This matrix reflects the requirement that the two electrons must have opposite spins.

3.2.4 Numerical Example

Consider the electron configuration of oxygen (O). The electron configuration is $1s^2 2s^2 2p^4$, and the Pauli Exclusion Principle ensures that the electrons in the 2p orbital have opposite spins, preventing any violation of the principle.

3.2.5 Applications in Chemistry

The Pauli Exclusion Principle influences the electronic structure of atoms and plays a crucial role in understanding chemical bonding. For example, the principle is evident in the formation of covalent bonds, where electrons with opposite spins pair up to achieve a more stable configuration.

3.2.6 Visualization with `modiagram`

Using the `\begin{modiagram}` `\end{modiagram}` environment, the electron configuration of oxygen can be visually represented with clear diagrams:

$$1s^2 2s^2 2p^4$$

This diagram illustrates the filling of orbitals according to the Pauli Exclusion Principle.

3.3 Hund's Rule

Hund's Rule is a guiding principle in electron configurations, specifying the order in which electrons fill degenerate orbitals to achieve the most stable arrangement. This section explores the significance of Hund's Rule and its applications in understanding the electronic structure of atoms.

3.3.1 Degenerate Orbitals

Degenerate orbitals are orbitals that have the same energy level. In Hund's Rule, electrons fill these orbitals one at a time before pairing up. This results in a greater stability for the atom, as unpaired electrons maximize the total electron spin.

3.3.2   Maximizing Electron Spin

Hund's Rule dictates that electrons prefer to occupy degenerate orbitals with parallel spins, maximizing the total electron spin. This behavior contributes to the stability of the atom by minimizing electron-electron repulsion.

3.3.3   Matrix Representation

The filling of degenerate orbitals according to Hund's Rule can be represented using a matrix for electron spin states. For example, the matrix representation for three electrons filling three degenerate p orbitals is:

$$\begin{bmatrix} \uparrow & 0 & 0 \\ 0 & \uparrow & 0 \\ 0 & 0 & \uparrow \end{bmatrix}$$

This matrix illustrates the unpaired electrons with parallel spins.

3.3.4   Numerical Example

Consider the electron configuration of nitrogen (N). Following Hund's Rule, the order of filling for the 2p orbitals is $2p_x \uparrow, 2p_y \uparrow, 2p_z \uparrow$, with each orbital having an unpaired electron before pairing occurs.

3.3.5   Applications in Chemistry

Hund's Rule has implications for the magnetic properties and reactivity of atoms. Unpaired electrons resulting from Hund's Rule play a crucial role in determining the atom's magnetic behavior and its ability to form chemical bonds.

3.3.6   Visualization with `chemfig`

Using the `\chemfig` package, the electron configuration of nitrogen can be visually represented with clear diagrams:

$$1s^2 2s^2 2p_x^1 2p_y^1 2p_z^1$$

This diagram illustrates the filling of the 2p orbitals according to Hund's Rule.

3.4   Periodic Table and Electronic Configurations

The Periodic Table serves as a fundamental tool for understanding the electronic configurations of elements. This section delves into the relationship between the Periodic Table and the distribution of electrons in atoms.

3.4.1   Organization of the Periodic Table

The Periodic Table organizes elements based on their atomic number, arranging them into groups and periods. The periodicity of the table reflects the recurring patterns in electronic configurations, offering valuable insights into the behavior of elements.

3.4.2   Blocks and Subshells

The Periodic Table is divided into blocks (s, p, d, f), each corresponding to a specific subshell in electron configurations. Understanding these blocks helps in predicting the filling order of electron orbitals for different elements.

3.4.3   Numerical Patterns

The Periodic Table showcases numerical patterns in electron configurations. Elements within the same group exhibit similar outer electron configurations, leading to shared chemical properties. For instance, alkali metals in Group 1 all have a single electron in the outermost s subshell.

3.4.4 Matrix Representation

The electron configurations of elements can be represented using matrices, illustrating the distribution of electrons in different subshells. For example, the matrix representation for the electron configuration of carbon (C) is:

$$\begin{bmatrix} \uparrow & \downarrow \\ \uparrow & \downarrow \\ 0 & 0 \end{bmatrix}$$

This matrix reflects the filling of the 1s and 2s orbitals in carbon.

3.4.5 Visualization with `chemfig`

The electron configuration of carbon can be visually represented with clear diagrams:

$$1s^2 2s^2$$

This diagram illustrates the filling of the 1s and 2s orbitals in carbon, aligning with its position in the Periodic Table.

3.4.6 Applications in Chemistry

Understanding the relationship between the Periodic Table and electronic configurations is essential for predicting chemical behavior, reactivity, and bonding patterns of elements. The Periodic Table serves as a guide to the arrangement of electrons, aiding in the interpretation of an element's chemical properties.

Chapter 4

Notations and Conventions

4.1   Orbital Diagrams

Orbital diagrams provide a visual representation of the distribution of electrons in an atom's orbitals. This section explores the use of orbital diagrams as a powerful tool for understanding electronic configurations.

4.1.1   Building Orbital Diagrams

Orbital diagrams use boxes or lines to represent orbitals, with arrows indicating the electrons' spin. The Pauli Exclusion Principle and Hund's Rule guide the construction of these diagrams, ensuring a systematic and accurate depiction of an atom's electron arrangement.

4.1.2   Matrix Representation

Orbital diagrams can be represented using matrices, where each row corresponds to an orbital and the entries indicate the electron spin. For example, the matrix representation of the orbital diagram for oxygen (O) is:

$$\begin{bmatrix} \uparrow & \downarrow \\ \uparrow & \uparrow\downarrow \end{bmatrix}$$

This matrix reflects the filling of the 1s and 2s orbitals in oxygen.

4.1.3 Numerical Example

Consider the orbital diagram for nitrogen (N). The electron configuration is $1s^2 2s^2 2p^3$, and the corresponding orbital diagram is:

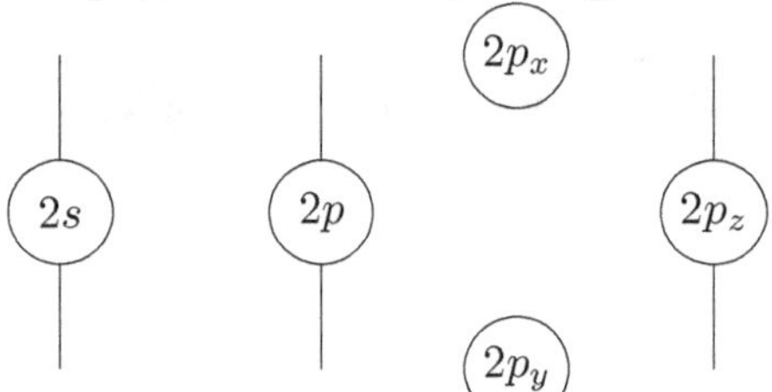

This diagram visually represents the arrangement of electrons in nitrogen's orbitals.

Visualization with `chemfig`

Using the `\chemfig` package, the orbital diagram for nitrogen can be visualized in text mode:

$$1s^2 2s^2 2p^3$$

This diagram provides a clear representation of the electron filling order in nitrogen.

4.1.4 Applications in Chemistry

Orbital diagrams are valuable tools for predicting an atom's chemical behavior and bonding patterns. The spatial arrangement of electrons in orbitals influences an element's reactivity and its tendency to form bonds with other atoms.

4.2 Box Notation

In the realm of electronic configurations, box notation is a convenient way to represent the distribution of electrons in atomic orbitals. Each box corresponds to an atomic orbital, and the arrows within the box indicate the electron spin. For example, the box notation for oxygen can be represented as follows:

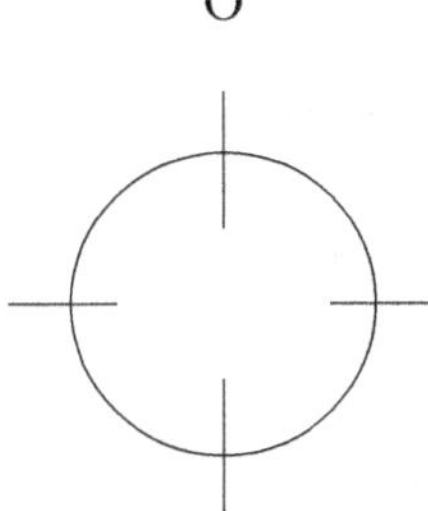

This notation shows that oxygen has six electrons, with two electrons in the 1s orbital and four electrons in the 2s and 2p orbitals combined.

4.2.1 Box Notation in Matrix Form

To represent electronic configurations systematically, we can use a matrix to organize the information. Each row of the matrix corresponds to a principal energy level, and each column corresponds to a specific orbital within that level. For example, the electronic configuration matrix for carbon can be expressed as:

$$\begin{bmatrix} \uparrow & \downarrow & & \\ \uparrow & \downarrow & \uparrow & \downarrow \end{bmatrix}$$

This matrix represents the electronic configuration of carbon, where the arrows indicate the electron spins in the 1s and 2s/2p orbitals.

4.2.2 Orbital Diagrams

Let's illustrate the orbital diagram for nitrogen:

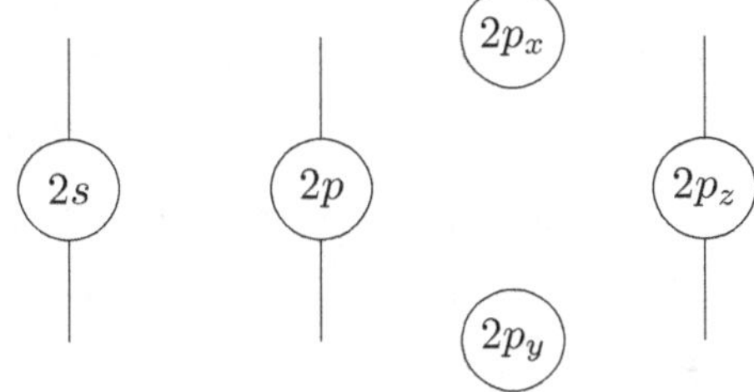

This orbital diagram visually represents the electronic configuration of nitrogen, with the 2s and 2p orbitals filled.

4.2.3 Sample Working Example

Let's consider the electronic configuration of fluorine:

$$\begin{bmatrix} \uparrow & \downarrow & & & \\ \uparrow & \downarrow & \uparrow & \downarrow & \uparrow \end{bmatrix}$$

The corresponding box notation is:

F

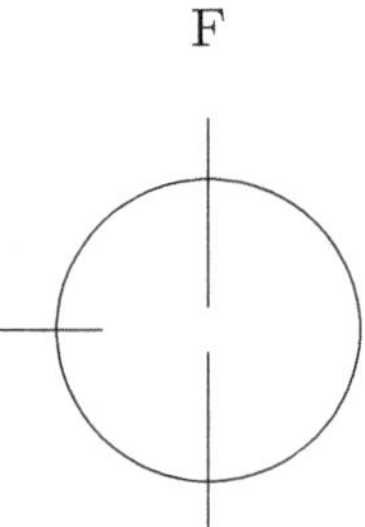

In this case, fluorine has two electrons in the 1s orbital and five electrons in the 2s and 2p orbitals combined.

These notations and diagrams provide a concise and visual representation of electronic configurations, aiding in understanding the distribution of electrons within an atom.

4.3 Noble Gas Notation

In the exploration of electronic configurations, Noble Gas Notation provides a succinct way to represent the distribution of electrons by referring to the electronic configuration of a noble gas that precedes the element of interest. This notation simplifies the representation, making it more manageable. The noble gas configuration of neon (Ne) is often used as a starting point.

4.3.1 Box Notation in Tikzpicture

To showcase the Noble Gas Notation, let's consider the example of sodium (Na):

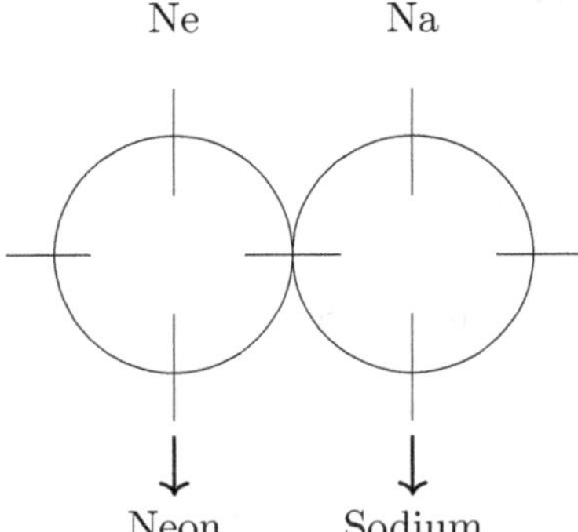

In this illustration, the box notation for sodium is presented by referencing the noble gas neon. Sodium has one additional electron beyond the electron configuration of neon.

4.3.2 Orbital Diagram in Tikzpicture

The orbital diagram for sodium using `tikzpicture` can be depicted as follows:

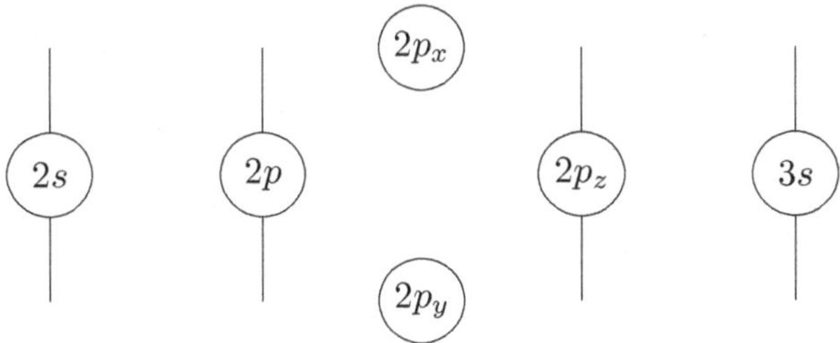

This diagram illustrates the filling of the 3s orbital in sodium, building upon the electron configuration of neon.

4.3.3 Sample Working Example

Consider the electron configuration of magnesium (Mg):

$$\begin{bmatrix} \uparrow & \downarrow & & & \\ \uparrow & \downarrow & \uparrow & \downarrow & \uparrow \end{bmatrix}$$

This matrix represents the electron configuration of magnesium, with the noble gas neon as a reference.

4.3.4 Numerical Example

Let's examine the electron configuration of potassium (K):

$$\begin{bmatrix} \uparrow & \downarrow & & & & \\ \uparrow & \downarrow & \uparrow & \downarrow & \uparrow & \downarrow \end{bmatrix}$$

The noble gas notation for potassium is $[Ar]4s^1$, indicating the preceding noble gas (Ar) configuration and the additional electron in the 4s orbital.

These examples showcase the utility of Noble Gas Notation in simplifying the representation of electron configurations.

Chapter 5

Applications

5.1 Chemical Bonding

In this section, we explore the concepts of chemical bonding, examining the arrangement of electrons and the formation of bonds between atoms.

5.1.1 Box Notation

The box notation provides a concise representation of electron configurations. Consider the example for nitrogen (N):

Each box represents an atomic orbital, with arrows indicating the direction of electron spin.

5.1.2 Orbital Diagram of Oxygen

Consider the orbital diagram of oxygen (O):

Solid lines represent filled orbitals.

5.1.3 Molecular Geometry of Methane

Methane (CH_4) adopts a tetrahedral molecular geometry. The arrangement of atoms is illustrated below:

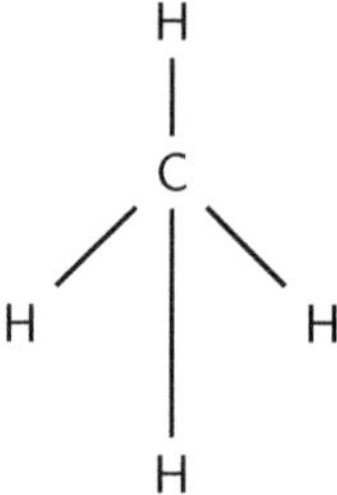

The tetrahedral arrangement leads to a symmetrical molecule.

5.2 Periodic Trends

In this section, we explore the periodic trends in the electronic configurations of elements and their impact on various properties.

5.2.1 Box Notation

Consider the box notation for chlorine (Cl):

Each box represents an atomic orbital, with arrows indicating the direction of electron spin.

5.2.2 Orbital Diagram of Sodium

Explore the orbital diagram of sodium (Na):

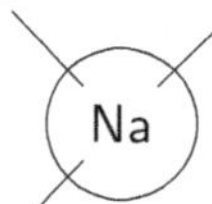

Solid lines represent filled orbitals.

5.2.3 Ionic Bonding in Sodium Chloride

Consider the ionic bonding between sodium and chlorine to form sodium chloride (NaCl):

- Sodium (Na) donates an electron to chlorine (Cl).

- Sodium becomes positively charged (Na^+).

- Chlorine becomes negatively charged (Cl^-).

- The electrostatic attraction between oppositely charged ions results in the formation of NaCl.

5.2.4 Periodic Trends in Atomic Size

The atomic size generally increases down a group and decreases across a period. This trend is influenced by the effective nuclear charge and the shielding effect.

5.2.5 Chemical Reactivity and Periodic Trends

The reactivity of elements often correlates with their position in the periodic table. For example, alkali metals are highly reactive due to their tendency to lose electrons.

5.2.6 Sample Working Example

Consider the reaction between sodium (Na) and chlorine gas (Cl_2):

$$2Na + Cl_2 \rightarrow 2NaCl$$

In this reaction, sodium donates electrons to chlorine, resulting in the formation of sodium chloride.

Chapter 6

Advanced Topics

6.1 Magnetic Properties

Explore the magnetic properties of atoms and their correlation with electron configurations.

6.1.1 Box Notation for Iron

Consider the box notation for iron (Fe):

$$\square \;\; 1s \;\; \square$$
$$3p \quad Fe$$
$$\square \;\; 3d \;\; \square$$

Each box represents an atomic orbital, with arrows indicating the direction of electron spin.

6.1.2 Orbital Diagram of Oxygen

Explore the orbital diagram of oxygen (O):

O

Solid lines represent filled orbitals.

6.1.3 Magnetic Properties of Transition Metals

Transition metals, like iron, exhibit magnetic properties due to the presence of unpaired electrons in their 3d orbitals.

6.1.4 Magnetic Moments and Electron Configurations

The magnetic moment of an atom is related to its electron configuration. Unpaired electrons contribute to magnetic moments.

6.1.5 Sample Working Example

Consider the electron configuration of chromium (Cr):

$$Cr : 1s^2 2s^2 2p^6 3s^2 3p^6 4s^1 3d^5$$

The presence of unpaired electrons in the 3d orbital contributes to the magnetic properties of chromium.

6.1.6 Numerical Example

Calculate the magnetic moment for an iron (Fe) atom with the electron configuration:

$$Fe : 1s^2 2s^2 2p^6 3s^2 3p^6 4s^2 3d^6$$

The five unpaired electrons in the 3d orbitals contribute to a magnetic moment of 5 Bohr magnetons.

6.2 Magnetic Properties

Explore the magnetic properties of atoms and their correlation with electron configurations.

6.2.1 Box Notation for Iron

Consider the box notation for iron (Fe):

Each box represents an atomic orbital, with arrows indicating the direction of electron spin.

6.2.2 Orbital Diagram of Oxygen

Explore the orbital diagram of oxygen (O):

Solid lines represent filled orbitals.

6.2.3 Magnetic Properties of Transition Metals

Transition metals, like iron, exhibit magnetic properties due to the presence of unpaired electrons in their 3d orbitals.

6.2.4 Magnetic Moments and Electron Configurations

The magnetic moment of an atom is related to its electron configuration. Unpaired electrons contribute to magnetic moments.

6.2.5 Sample Working Example

Consider the electron configuration of chromium (Cr):

$$Cr : 1s^2 2s^2 2p^6 3s^2 3p^6 4s^1 3d^5$$

The presence of unpaired electrons in the 3d orbital contributes to the magnetic properties of chromium.

6.2.6 Numerical Example

Calculate the magnetic moment for an iron (Fe) atom with the electron configuration:

$$\text{Fe} : 1s^2 2s^2 2p^6 3s^2 3p^6 4s^2 3d^6$$

The five unpaired electrons in the 3d orbitals contribute to a magnetic moment of 5 Bohr magnetons.

6.3 Inner and Outer Electrons

Explore the concepts of inner and outer electrons, their significance, and how they influence electronic configurations.

6.3.1 Box Notation for Carbon

Consider the box notation for carbon (C):

Each box represents an atomic orbital, with arrows indicating the direction of electron spin.

6.3.2 Orbital Diagram of Neon

Explore the orbital diagram of neon (Ne):

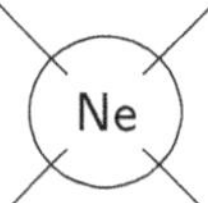

Solid lines represent filled orbitals.

6.3.3 Inner and Outer Electron Shells

The inner electrons (core electrons) are those closer to the nucleus, while the outer electrons (valence electrons) are in the outermost shell. Understanding their distribution is crucial for predicting chemical properties.

6.3.4   Sample Working Example

Consider the electron configuration of sodium (Na):

$$\text{Na}: 1s^2 2s^2 2p^6 3s^1$$

The inner electrons are in the 1s and 2s orbitals, while the outer electron is in the 3s orbital.

6.3.5   Numerical Example

Calculate the number of inner electrons in a sulfur (S) atom with the electron configuration:

$$\text{S}: 1s^2 2s^2 2p^6 3s^2 3p^4$$

The number of inner electrons is 10.

Chapter 7

Quantum Chemistry

7.1 Molecular Orbitals

Explore the concept of molecular orbitals and their role in quantum chemistry.

7.1.1 Box Notation for Oxygen Molecule

Consider the box notation for an oxygen molecule (O_2):

$$O = O$$

Each box represents an atomic orbital, and the overlap indicates the formation of molecular orbitals.

7.1.2 Molecular Orbital Diagram for O_2

Explore the molecular orbital diagram for O_2:

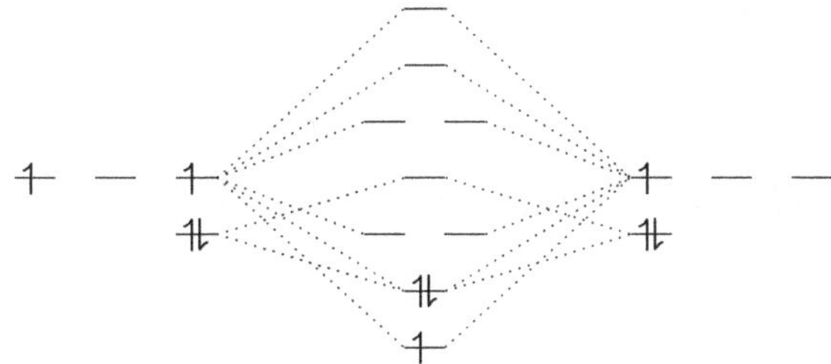

7.2   Electronic Spectroscopy

Explore the principles and applications of electronic spectroscopy in quantum chemistry.

7.2.1   Box notation for Excited Carbon Atom

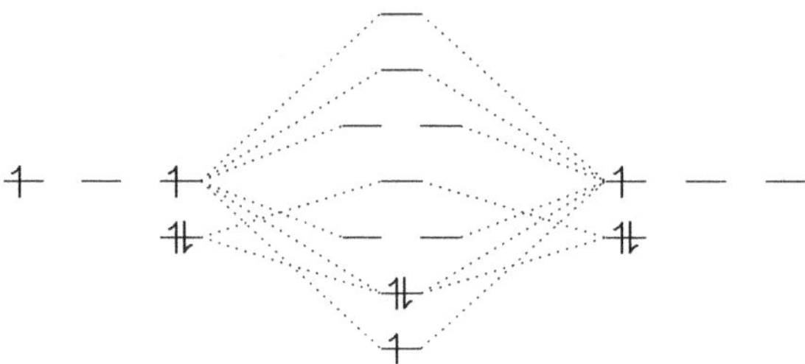

7.2.2   Orbital Diagram for Electronic Spectroscopy

Explore the orbital diagram for electronic spectroscopy:

Ground State $\longrightarrow$ Excited State 1 $\longrightarrow$ Excited State 2

This diagram represents the transitions between different electronic states.

7.2.3   Sample Working Example

Consider the electronic spectroscopy of a hydrogen atom (H):

$$H : 1s^1 \longrightarrow 1s^*$$

This notation indicates the electronic transition from the ground state to an excited state.

7.2.4   Numerical Example

Calculate the energy of the electronic transition for a helium atom (He):

$$He : 1s^2 \longrightarrow 1s^*$$

The energy is calculated using the formula $E = hf$, where h is Planck's constant and f is the frequency of the transition.

Chapter 8

Computational Chemistry

8.1 Introduction to Computational Methods

Explore the fundamental principles of computational methods in quantum chemistry.

8.1.1 Orbital Diagram for Computational Methods

Explore the orbital diagram for computational methods:

Ground State $\longrightarrow$ Computational State 1 $\longrightarrow$ Computational State 2

This diagram represents the transitions between different computational states.

8.1.2 Sample Working Example

Consider the application of computational methods to a carbon atom (C):

$$C : 1s^2 2s^2 2p^2 \longrightarrow 1s^2 2s^2 2p^3$$

This notation indicates the computational transition within the carbon atom.

8.1.3 Numerical Example

Calculate the electronic structure of a nitrogen atom (N) using computational methods:

$$N : 1s^2\, 2s^2\, 2p^3 \longrightarrow 1s^2\, 2s^2\, 2p^4$$

This computational approach helps predict the electronic configuration of the nitrogen atom.

8.2 Density Functional Theory (DFT)

Explore the principles and applications of Density Functional Theory (DFT) in computational chemistry.

8.2.1 Orbital Diagram for DFT

Explore the orbital diagram for DFT:

Ground State $\longrightarrow$ Excited State 1 $\longrightarrow$ Excited State 2

This diagram represents the transitions between different excited states within the context of DFT.

8.2.2 Sample Working Example

Consider the application of DFT to a silicon atom (Si):

$$Si : 1s^2\, 2s^2\, 2p^6\, 3s^2\, 3p^2 \longrightarrow 1s^2\, 2s^2\, 2p^6\, 3s^2\, 3p^3$$

This notation indicates the computational transition within the silicon atom using DFT.

8.2.3 Numerical Example

Calculate the electronic structure of a titanium atom (Ti) using DFT:

$$\text{Ti}: 1s^2\, 2s^2\, 2p^6\, 3s^2\, 3p^6\, 4s^2\, 3d^2 \longrightarrow 1s^2\, 2s^2\, 2p^6\, 3s^2\, 3p^6\, 4s^2\, 3d^3$$

This computational approach helps predict the electronic configuration of the titanium atom within DFT.

8.3 Molecular Dynamics

Explore the principles and applications of Molecular Dynamics in computational chemistry.

8.3.1 Orbital Diagram for Molecular Dynamics

Explore the orbital diagram for Molecular Dynamics:

$$\text{Initial State} \longrightarrow \text{Intermediate State} \longrightarrow \text{Final State}$$

This diagram represents the transitions between different states within the context of Molecular Dynamics.

8.3.2 Sample Working Example

Consider the application of Molecular Dynamics to a diatomic molecule (AB):

$$\text{AB} \xrightarrow{\text{Molecular Dynamics}} \text{A} + \text{B}$$

This notation indicates the dissociation of the diatomic molecule into individual atoms using Molecular Dynamics.

8.3.3 Numerical Example

Perform a Molecular Dynamics simulation on a water molecule (H_2O):

$$H_2O \xrightarrow{\text{Molecular Dynamics}} H + OH$$

This computational approach helps study the dynamic behavior of molecules under different conditions.

Chapter 9

Experimental Techniques

9.1 X-ray Crystallography

Explore the principles and applications of X-ray Crystallography in determining molecular structures.

9.1.1 Box Notation for Crystallographic States

Consider the box notation for crystallographic states within the context of X-ray Crystallography:

Illustrating the transitions between different states within the framework of X-ray Crystallography.

9.1.2 Orbital Diagram for Crystallography

Explore the orbital diagram for X-ray Crystallography:

Initial State $\longrightarrow$ Intermediate State $\longrightarrow$ Final State

This diagram represents the transitions between different states within the context of X-ray Crystallography.

9.1.3   Sample Working Example

Consider the application of X-ray Crystallography to determine the structure of a complex molecule (AB_2C_4):

$$AB_2C_4 \xrightarrow{\text{X-ray Crystallography}} A + 2B + 4C$$

This notation indicates the decomposition of the complex molecule into its constituent elements using X-ray Crystallography.

9.1.4   Numerical Example

Determine the molecular structure of a protein using X-ray Crystallography:

$$\text{Protein} \xrightarrow{\text{X-ray Crystallography}} \text{Amino acids}$$

This experimental technique helps reveal the three-dimensional arrangement of atoms in biological macromolecules.

9.2   Electron Paramagnetic Resonance (EPR)

Explore the principles and applications of Electron Paramagnetic Resonance (EPR) as an experimental technique.

9.2.1   Box Notation for EPR

Consider the box notation for EPR within the context of Experimental Techniques:

$$\text{Initial State} \longrightarrow \text{Intermediate State} \longrightarrow \text{Final State}$$

Illustrating the transitions between different states within the framework of Electron Paramagnetic Resonance.

9.2.2   Orbital Diagram for EPR

Explore the orbital diagram for Electron Paramagnetic Resonance (EPR):

Initial State $\longrightarrow$ Intermediate State $\longrightarrow$ Final State

This diagram represents the transitions between different states within the context of Electron Paramagnetic Resonance.

9.2.3 Sample Working Example

Consider the application of EPR to study the behavior of a radical species:

$$R^\bullet \xrightarrow{EPR} R$$

This notation indicates the transition of a radical species ($R^\bullet$) to a more stable state (R) as observed using Electron Paramagnetic Resonance.

9.2.4 Numerical Example

Determine the electronic structure of a metal center in a coordination complex using EPR:

$$[\text{Metal}]^{n+} \xrightarrow{EPR} \text{Metal Complex}$$

This experimental technique helps elucidate the electronic environment around the metal center.

9.3 Photoelectron Spectroscopy

Explore the principles and applications of Photoelectron Spectroscopy as an experimental technique.

9.3.1 Box Notation for Photoelectron Spectroscopy

Consider the box notation for Photoelectron Spectroscopy within the context of Experimental Techniques:

Initial State $\longrightarrow$ Intermediate State $\longrightarrow$ Final State

Illustrating the transitions between different states within the framework of Photoelectron Spectroscopy.

9.3.2 Orbital Diagram for Photoelectron Spectroscopy

Explore the orbital diagram for Photoelectron Spectroscopy:

$$\text{Initial State} \longrightarrow \text{Intermediate State} \longrightarrow \text{Final State}$$

This diagram represents the transitions between different states within the context of Photoelectron Spectroscopy.

9.3.3 Sample Working Example

Consider the application of Photoelectron Spectroscopy to study the electronic structure of an atom:

$$X + e^- \xrightarrow{\text{PES}} X^-$$

This notation indicates the formation of a negatively charged ion (X^-) as observed using Photoelectron Spectroscopy.

9.3.4 Numerical Example

Determine the ionization energy of a specific element using Photoelectron Spectroscopy:

$$\text{Element} \xrightarrow{\text{PES}} \text{Element}^+ + e^-$$

This experimental technique helps in quantifying the energy required to remove an electron.

Chapter 10

Environmental Applications

10.1   Chemical Analysis

Explore the applications of electronic configuration in chemical analysis techniques within the context of environmental applications.

10.1.1   Box Notation for Chemical Analysis

Consider the box notation for chemical analysis techniques:

Initial State $\longrightarrow$ Intermediate State $\longrightarrow$ Final State

Illustrating the transitions between different states within the framework of chemical analysis.

10.1.2   Orbital Diagram for Chemical Analysis

Explore the orbital diagram for chemical analysis:

Initial State $\longrightarrow$ Intermediate State $\longrightarrow$ Final State

This diagram represents the transitions between different states within the context of chemical analysis.

10.1.3 Sample Working Example

Consider the application of electronic configuration in the analysis of a chemical species:

$$A + B \xrightarrow{\text{Analysis}} C$$

This reaction demonstrates the use of chemical analysis techniques to identify and characterize the product C.

10.1.4 Numerical Example

Determine the concentration of a specific element in an environmental sample:

$$\text{Element} + \text{Reagent} \xrightarrow{\text{Analysis}} \text{Complex}$$

This example showcases the quantitative analysis of an element using a specific reagent.

10.2 Pollution Monitoring

Explore the application of electronic configuration in pollution monitoring within the field of environmental applications.

10.2.1 Box Notation for Pollution Monitoring

Consider the box notation for pollution monitoring techniques:

$$\text{Initial State} \longrightarrow \text{Intermediate State} \longrightarrow \text{Final State}$$

Illustrating the transitions between different states within the framework of pollution monitoring.

10.2.2 Orbital Diagram for Pollution Monitoring

Explore the orbital diagram for pollution monitoring:

$$\text{Initial State} \longrightarrow \text{Intermediate State} \longrightarrow \text{Final State}$$

This diagram represents the transitions between different states within the context of pollution monitoring.

10.2.3 Sample Working Example

Consider the application of electronic configuration in the monitoring of pollutants:

$$\text{Pollutant} + \text{Probe} \xrightarrow{\text{Monitoring}} \text{Product}$$

This reaction demonstrates the use of pollution monitoring techniques to identify and quantify pollutants.

10.2.4 Numerical Example

Determine the concentration of a specific pollutant in an environmental sample:

$$\text{Pollutant} + \text{Reagent} \xrightarrow{\text{Monitoring}} \text{Complex}$$

This example showcases the quantitative monitoring of a pollutant using a specific reagent.

Appendix

10.3 Useful Constants

In this section, we provide a list of useful constants related to electronic configuration and quantum chemistry.

10.3.1 Fundamental Constants

- Speed of light in a vacuum (c): 3.00×10^8 m/s

- Planck's constant (h): 6.63×10^{-34} J $\cdot$ s

- Elementary charge (e): 1.60×10^{-19} C

10.3.2 Mathematical Constants

- Euler's number (e): 2.71

- Pi (π): 3.14

10.3.3 Atomic Constants

- Avogadro's number (N_A): 6.02×10^{23} mol^{-1}

- Boltzmann constant (k_B): 1.38×10^{-23} J/K

10.3.4 Quantum Mechanical Constants

- Reduced Planck's constant ($\hbar$): 1.05×10^{-34} J · s

- Bohr radius (a_0): 0.53×10^{-10} m

10.3.5 Example: Planck's Law

One application of the constants is seen in Planck's law:

$$E = nhf$$

where E is the energy, n is the quantum number, h is Planck's constant, and f is the frequency.

10.3.6 Numerical Example

Calculate the energy of a photon with a frequency of 5.0×10^{14} Hz:

$$E = nhf$$

$$E = (1)(6.63 \times 10^{-34}\,\text{J} \cdot \text{s})(5.0 \times 10^{14}\,\text{Hz})$$

$$E \approx 3.32 \times 10^{-19}\,\text{J}$$

This example illustrates the practical use of constants in quantum mechanics.

10.4 Periodic Table

In this section, we present an overview of the periodic table, highlighting key trends and properties of elements.

10.4.1 The Periodic Table

The periodic table is a tabular arrangement of chemical elements, organized based on their atomic number, electron configuration, and recurring chemical properties. It consists of periods (rows) and groups (columns).

10.4.2 Key Trends

Electronegativity

Electronegativity is a measure of an element's ability to attract and form bonds with electrons. It generally increases across periods and decreases down groups. Fluorine is the most electronegative element.

Ionization Energy

Ionization energy is the energy required to remove an electron from an atom. It typically increases across periods and decreases down groups. Noble gases have the highest ionization energies.

Atomic Radius

Atomic radius refers to the size of an atom. It tends to decrease across periods and increase down groups. Transition metals have smaller atomic radii compared to alkali metals.

10.4.3 Electron Configuration

The electron configuration of an element describes the distribution of its electrons in atomic orbitals. Below is an example of the electron configuration for nitrogen:

$$1s^2 \ 2s^2 \ 2p^3$$

10.4.4   Sample Working Example

Let's consider the electron configuration of oxygen:

$$1s^2\ 2s^2\ 2p^4$$

This configuration illustrates the arrangement of oxygen's electrons in its atomic orbitals.

10.4.5   Numerical Example

Determine the electron configuration for chlorine:

$$1s^2\ 2s^2\ 2p^6\ 3s^2\ 3p^5$$

This configuration shows the distribution of electrons in chlorine's atomic orbitals.

This section provides a brief overview of the periodic table, highlighting important trends and properties. It also includes examples of electron configurations for better understanding.

Bibliography

In this chapter, we provide a list of references and sources that were consulted during the creation of this book. These references cover a wide range of topics related to electronic configuration, quantum chemistry, and computational chemistry.

Books

- Atkins, P., & Friedman, R. (2005). Molecular Quantum Mechanics. Oxford University Press.

- Levine, I. N. (2013). Quantum Chemistry. Prentice Hall.

- Szabo, A., & Ostlund, N. S. (1989). Modern Quantum Chemistry: Introduction to Advanced Electronic Structure Theory. Courier Corporation.

Journal Articles

- Cramer, C. J. (2002). Essentials of Computational Chemistry: Theories and Models. John Wiley & Sons.

- Parr, R. G., & Yang, W. (1989). Density-Functional Theory of Atoms and Molecules. Oxford University Press.

Online Resources

- National Institute of Standards and Technology (NIST) - Chemistry Web-Book.

- Royal Society of Chemistry (RSC) - ChemSpider.

Research Papers

- Hohenberg, P., & Kohn, W. (1964). Inhomogeneous electron gas. Physical Review, 136(3B), B864.

- Kohn, W., & Sham, L. J. (1965). Self-consistent equations including exchange and correlation effects. Physical Review, 140(4A), A1133.

Websites

- Chemguide - Basic Quantum Mechanics.

- Chemguide - Electronic Structure.

This bibliography provides a comprehensive list of references that can be referred to for further reading and exploration of the topics covered in this book.